Large lakes of Europe:

9. English Lake District
   (e.g. Windermere, Ullswater)
10. Scottish Lochs
   (e.g. Ness, Lomond)
11. Neagh
12. Italian Piedmont Lakes
   (e.g. Como, Maggiori, Garda)
13. Glint Line lakes
   (e.g. Storsjon)
14. Bodensee
15. Geneva
16. Balaton
17. Vanern
18. Vattern
19. Finnish lakeland
20. Ladozhskoye

Large lakes of Asia:

21. Baikal
22. Aral Sea
23. Black Sea
24. Caspian Sea
25. Dead Sea
26. Poyang Hu

Large lakes of Australasia:

27. Eyre
28. Torrens
29. Gardiner
30. Mackay
31. Southern Alps lakes
   (e.g. Wakatipu, Te Anau, Wanaka)

FACTS ABOUT LAKES

A lake is any land-locked hollow filled with fresh or salt water. The largest lakes are often called seas, the smallest may be called ponds.

The Caspian Sea, in Central Asia, is the world's largest lake. It occupies over 148,000 square miles and contains 21,000 cubic miles of water. The surface of this extraordinary lake is 100 ft *below* sea level and yet it is over 3300 ft deep. Because no rivers drain from it, the salt in the water brought by the rivers cannot be flushed away and the lake is as salty as the world's oceans.

Lake Superior, in North America, is the largest lake containing entirely fresh water. It covers over 31,800 square miles. Its surface is nearly 600 ft above sea level and it drains to the Atlantic Ocean through the St. Lawrence river.

Lake Baikal, in Russia, covers a smaller area (12,000 square miles) but it is deeper than Lake Superior. In fact it holds 5,500 cubic miles of water, the world's greatest volume of fresh lake water. Lake Baikal is also the world's deepest lake, with its bottom 6300 ft below the surface, and therefore nearly 5000 ft below sea level.

Most of the world's lakes have been produced by the scouring action of ice or the result of being blocked by material dropped by ice sheets. Glacial lakes are the world's highest and, for example, in the Himalayas they can be found 19,000 ft above sea level. The world's lowest lake surface is the Dead Sea at 1200 ft below sea level. The greatest number of lakes found in any one place is in southern Finland. Here the thousands of small lakes are separated by ridges of glacial materials.

Grolier Educational Corporation
SHERMAN TURNPIKE, DANBURY, CONNECTICUT 06816

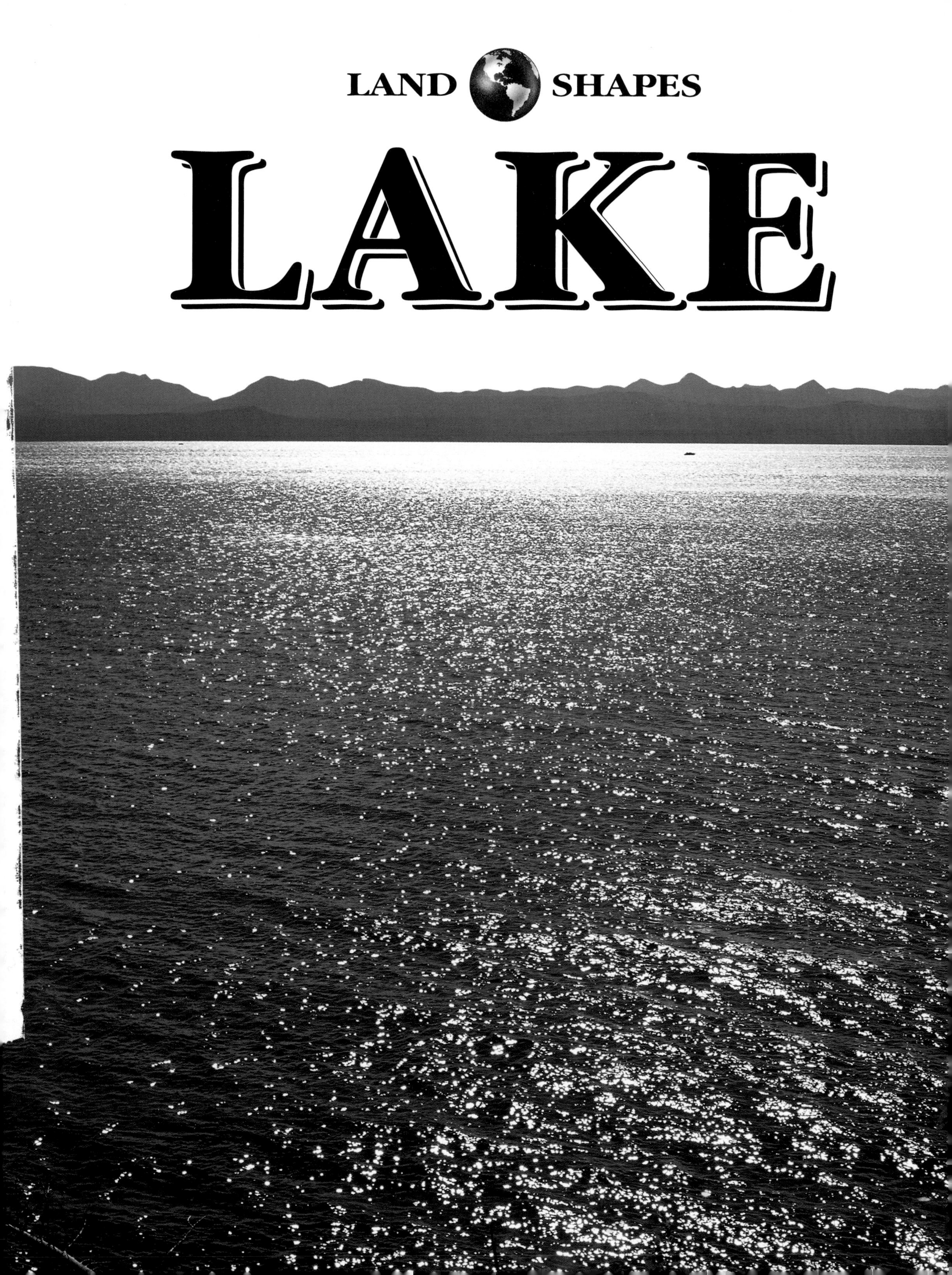

LAND SHAPES

LAKE

Author
Brian Knapp, BSc, PhD
Art Director
Duncan McCrae, BSc
Editor
Rita Owen
Illustrators
David Hardy and David Woodroffe
Print consultants
Landmark Production Consultants Ltd
Printed and bound in Hong Kong
Designed and produced by
EARTHSCAPE EDITIONS

First published in the USA in 1993 by
GROLIER EDUCATIONAL CORPORATION,
Sherman Turnpike, Danbury, CT 06816

Copyright © 1992
Atlantic Europe Publishing Company Limited

Library of Congress #92–072045

Cataloging information may be obtained
directly from Grolier Educational Corporation

Title ISBN 0–7172–7187–0

Set ISBN 0–7172–7176–5

All rights reserved. No part of this publication may be reproduced, stored in a retrieval system, or transmitted in any form or by any means otherwise, without prior permission in writing of the publisher, nor be otherwise circulated in any form of binding or cover other than that in which it is published and without a similar condition including this condition being imposed on the subsequent purchaser.

Acknowledgements. The publishers would like to thank the following: Horizon Air and Redlands County Primary School.

Picture credits. All photographs from the Earthscape Editions photographic library except the following (t=top, b=bottom, l=left, r=right): Hutchison Library 25; NASA 35tl, 35bl; Sirrel Young 10l; ZEFA 8/9, 18/19, 19br, 24, 32/33, 33tr, 34bl.

Cover picture: Derwentwater, Lake District, UK.
Inside back cover picture: St Mary Lake, Waterton/Glacier International Peace Park, Canada/USA border.

In this book you will find some words that have been shown in **bold** type. There is a full explanation of each of these words on page 36.

On some pages you will find experiments that you might like to try for yourself. They have been put in a blue box like this.

In this book mi means miles and ft means feet.

These people appear on a number of pages to help you to know the size of some landshapes.

CHILDREN'S ROOM

CONTENTS

JR551.4
K727x
V.12

Introduction

A lake is an area of still water that is closed off from the sea. Lakes can be wide and shallow or narrow and deep, they can even exist below sea level. There are lakes in the midst of deserts that are more salty than the sea and lakes in mountain ranges, so crystal clear that fish can be seen swimming tens of feet below the surface.

Most lakes form in basins that are the result of glaciers scouring the land during the **Ice Age**. Others were formed when rivers were blocked by material left behind by ice sheets. The biggest lakes, however, occur where the land has sunk because of movements in the Earth's crust that created huge trench-like depressions called **rift valleys**. The world's largest lake lies in such a trench and it is so big it has been named the Caspian Sea.

In terms of Earth history, lakes do not have very long lives. Rivers bring with them mud, sand and pebbles that gradually fill the lakes in. In this book you can read about the many ways lakes have been created and the way they are eventually changed back into dry land. Just turn to a page to enjoy the landshapes of lakes.

This is Lake Nakuru in East Africa. You can tell that it is a shallow lake because flamingoes are able to walk in it searching for their insect food in the lake muds.

Take care near lakes

It is fun to visit lakes to see the landshapes described in this book for yourself. But never take a boat out on a lake without an adult. Lakes can be dangerous places for the unwary and deaths have occurred because people have capsized in the water.

Chapter 1:
How lakes work

A lake is born

There are many natural processes that prevent rivers from flowing to the sea. For example, a volcano may erupt and send a tongue of lava into a river valley, sealing one end, or an earthquake may cause a part of the land surface to drop down.

Landslides are the most common of the sudden lake-making processes. A landslide may release millions of tons of soil and rock and in minutes make a natural dam across a valley. This is how Lake Thistle was formed, and over the next few weeks people were able to see the lake grow.

Why Lake Thistle was formed

The spring of 1985 had been very wet in the Wasatch Mountains of Utah, USA, and there was still much snow on the mountains.

As the rain fell and the snow melted, enormous amounts of water soaked into the steeply-sloping soils and surface rocks.

Eventually so much water was in the soil that it lost its grip on the rock and began to slide. A great tongue of soil and loose rock slid down the mountain and within minutes filled the valley bottom.

The picture on the left shows the area where the landslide occurred, and when the lake was at its greatest.

Why the lake dried up

Lake Thistle only existed for a few months. If it had been left untouched Lake Thistle would still be there today. But people were afraid that the landslide might flow away allowing a huge amount of lake water to spill down the valley and destroy yet more towns. A small diversion tunnel, therefore, was built and the lake was drained.

Several houses were drowned by Lake Thistle and when the lake was drained they were found to be half filled with mud as the picture below reveals. This shows quite dramatically how much sediment is brought into a lake by a flooding river in such a short period of time.

A flat lake bed was formed in just a few short weeks. This picture shows what it was like soon after the lake was drained.

Landshapes of lakes

Many of the shapes and processes that you can see around oceans can be seen, in miniature, around lakes. For example, some lakes have beaches and cliffs, waves and currents. Rivers bring in material and eventually build up **deltas**.

Lakes are often fringed with plants that trap the fine materials suspended in the water, creating **swampy** or marshy areas. As the swamps grow outwards into the lake, so they change its shape.

As the seasons pass, the amount of water entering and leaving the lake will often change. Some lakes may dry out completely, in others the water level is simply lowered. The history of these changes can be seen in the detailed shape of a lake.

This is the outlet, where the water leaves the lake during times of high water level (see page 14).

This small lake occupies a hollow in the 'blanket' of debris dropped from an ancient ice sheet (see page 27).

The changes are easier to spot in small lakes and ponds rather than large lakes and seas. By walking round the edge of a lake like this it would be easy to see how the plants are trapping mud and making the lake ever smaller (see page 19).

The darkest blue color shows where the lake is deepest (see page 20).

During a storm, water flows off the land surface all around the lake, bringing in **sediment** which then settles at the lake edge (see page 16).

Here streams have flowed in and brought debris with them. It has settled in the lake and has built up almost to the surface (see page 16).

Sometimes artificial lakes – reservoirs – are lowered and old shorelines emerge. This is a good opportunity to see how much sediment is dropped on a 'lake' bed.

Why lakes swell and shrink

Unlike rivers, which cut their own channels, lakes simply inherit a land abandoned by ice, blocked by an obstacle or land that has sagged because of forces inside the Earth. But once formed, the gradual swelling and shrinking of a lake has a large effect on the way rivers flow and sediment is deposited.

Sources of water

A lake can gather water from many sources. Quite often the major source seems to be from rivers, direct rainfall or melting snow, but scientists have discovered that many lakes are fed with water that seeps through the soil of the banks or through the rock of the lake bed.

Where water flows out

Water spills out of the lake at its lowest point and will supply a stream or river. It is called the outflow point. If the outflow point is worn down the lake level will fall and the lake will shrink.

Because the outfall is broad, a rise in water level of just a few inches would allow very large quantities of water to leave the lake.

The broad outflow is the reason why there is so little obvious change in size and height in lakes even when rivers are in flood.

Slowly filling lakes

This river swells rapidly after each storm as it gathers water from the land. But even the water in this river is but a tiny drop compared to the huge amount stored in the distant lake.

Rivers bringing water to the mountain lake.

Bowl-shaped hollow scoured by a glacier and now filled by the lake.

This is the 'lip' of the ice-scoured hollow.

See why lake levels change so slowly

There is so much water stored in a lake that even rivers bringing down floodwaters after large storms may be absorbed into the lake.

You can see this effect by making a lake from a tilted tank and using a hose-pipe to play the part of a river.

Fill the tank with water and then turn the hose down so that water just flows in slowly. Mark the level of the water in the tank.

Now turn the hose on full to represent a river bringing flood waters into it and see what happens. You may be surprised at how long it takes for the level to rise; this shows the enormous storing power of the lake.

Mark the height the lake reaches with your 'flooding' river; it is likely to be only a little higher than for a small flow. Then turn the water off and see what happens.

Deltas

A delta is a fan-shaped piece of land that has been built out into the lake by a river. The two banks of the river are usually higher than the rest of land. They are called **levees** and they guide the water into the lake. Occasionally these levees are breeched because the river is carrying too much storm water. These breeches are called crevasses and they allow the water and sediment to flow sideways and widen the delta. In the main channel the coarsest material carried in the water is usually dropped as soon as it reaches the still waters of the lake, where it builds up into banks called shoals.

The part of the delta that you can see is, however, only a very small part of the real landshape, because the sides of a delta are gently sloping, stretching much further out into the lake and hidden under the water.

Deltas form quickly when they are supplied by debris-laden waters from a melting glacier. The strange turquoise color of the lake shown here is produced by tiny rock fragments still suspended in the water. This is typical of a lake that is still being fed by a melting glacier.

Where mud is trapped

A delta is the clearest evidence that the debris brought in by rivers is settling out and filling in the lake. But the finer debris does not settle so easily, and often drifts around the lake with the slightest currents until it is trapped by lakeside plants or eventually settles out on the lake bottom.

Lakes build their beds

In this book many types of lake are described, each of which is being filled in at a different rate. How quickly a lake fills in will depend on the amount of debris that is carried in by the rivers.

In nearly every case there will be times when rivers bring in more materials than others, for example during a flood, or in winter. Each period will be silently marked by the way the lake muds settle out – and just occasionally it is possible to see the layers which mark time just as surely as tree rings.

These muddy layers from an ancient lake bed were formed in a lake near to a glacier. Because the glacier released huge amounts of debris as it melted each summer, the 'summer' layers are quite thick. When the lake was frozen in winter little material settled and the bands are thin.

A layer deposited in summer.

A layer deposited in winter.

Trapped mud builds up on the lake bottom and makes the lake shallower. Where the water remains too deep for plants to grow it is impossible to gauge the depth of the lake, but when it reaches close to the surface plants are able to grow and form a swamp or marsh.

The picture on the right shows a swamp on Lake Titicaca where the rushes have trapped the mud around their roots and stems. Swamps and marshes expand quickly on lakes with gently sloping shores.

Investigating lake changes

As soon as a lake forms, important changes begin. Some, such as waves cutting cliffs around the margins, are easy to see; others, such as debris settling on the lake bottom and slowly making the lake shallower, must be investigated by 'plumbing' the depths.

Changes in the shore
If you have a map of the lake you can walk round the edge, marking where the lake is marshy (= shallow) and where there is a pebbly beach or a small cliff (= exposed). This pattern may help you to discover just how the shape of the lake is changing.

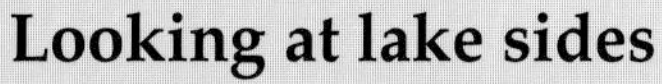

Looking at lake sides
Lake sides are usually covered with plants and they can be marshy. Carefully collect soil samples from around the lakeside plants. Stir a sample of it in a jar of water and see how quickly it settles. Are there many different sizes of material in the sample or just fine mud?

Beach ridges mark the changing lake levels.

Wave-cut cliff in soft lake-edge material.

Finding the shape of a lake edge and lake bed

An easy way of finding the real shape of a lake edge is to survey it using a plumb-line made from a piece of fishing line and a fishing rod.

Attach a weight to the end of the line (do not use lead or other materials that could be harmful to wildlife). Use a small piece of transparent sticky tape to tag the fishing line every 5 inches and using waterproof ink mark the value on each tag. The first tag should be placed 5 inches from the bottom of the weight.

Use the fishing rod to find out about the way the lake bank slopes. If you can find a jetty to walk out on this will give you an even greater distance for your survey. Measure how far you have hung the rod out over the water and make a note of the length of line you had to pay out by counting the tags as they sink below the water.

If an adult will take you out in a boat, the tagged fishing line can be used to find the deepest part of a lake. Ask them to row or drive slowly over the lake, then drop the weighted line from time to time until you find the deepest part of the lake.

Caution

Work from a boat should only be done under the supervision of an adult.

A lake edge that shows bays and headlands. Cliffs and rocky shores are found where the wave energy is greatest, probably where the winds drive waves onshore.

Chapter 2: Many types of lakes

Lakes in mountain hollows

Even the toughest rocks on Earth cannot withstand the scouring power of glaciers. During the Ice Age, as glaciers were fed with snow, they began to push their way down from mountain peaks creating valleys.

As the ice pushed forward, the underside plucked any loose and fractured rock from its bed, and, armed with these powerful tools, it began to grind at the landscape. The weakest rocks were soon deeply scoured, while the tougher ones were less easy to attack. So, gradually the weaker rocks became hollows and the tougher ones stood proud as raised ridges.

Later, when the ice melted and rivers began flowing again, the scoured hollows were filled in with water – in fact mountain lakes were born.

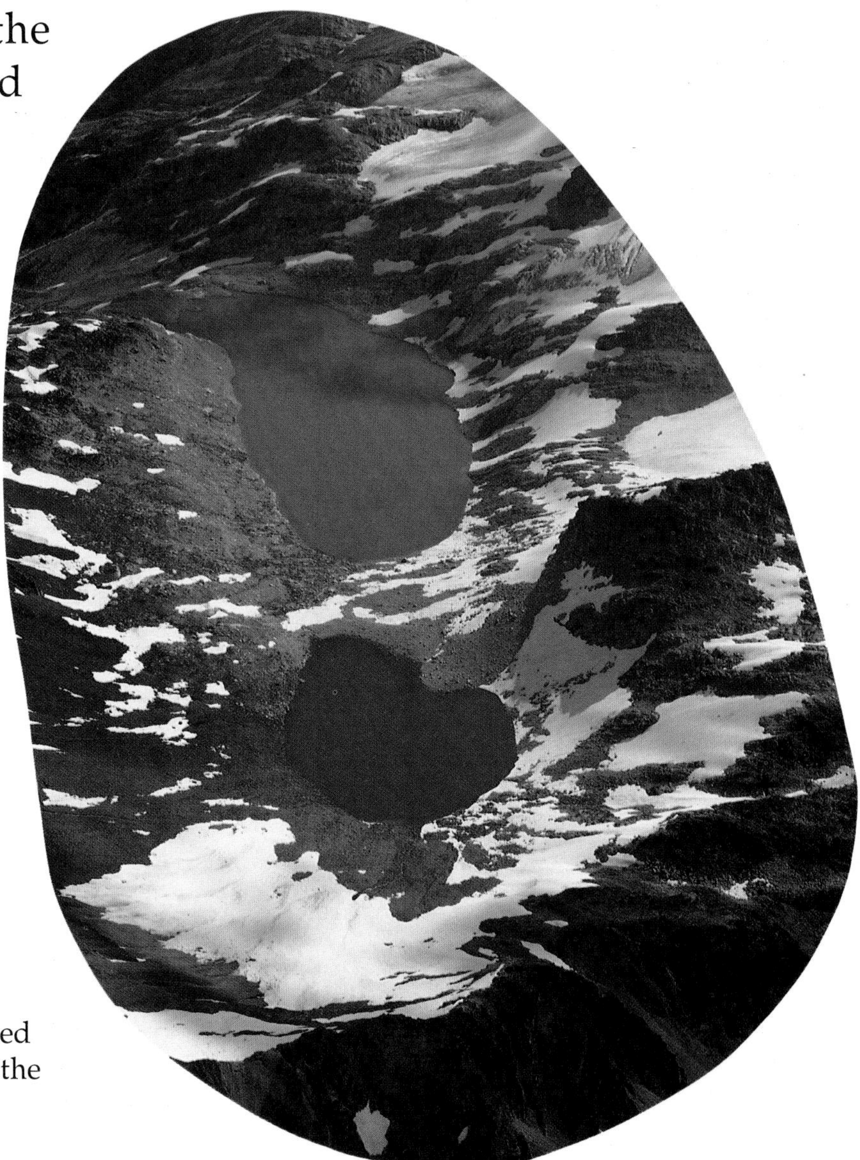

It is often helpful to look at landshapes from more than one direction. This picture was taken from above the mountain peaks and you can see the bowl shape of the upper lake and the moraine that helps to trap its waters. Notice, too, the way that a recent landslide has spilled debris into the lake, beginning the process of filling it in.

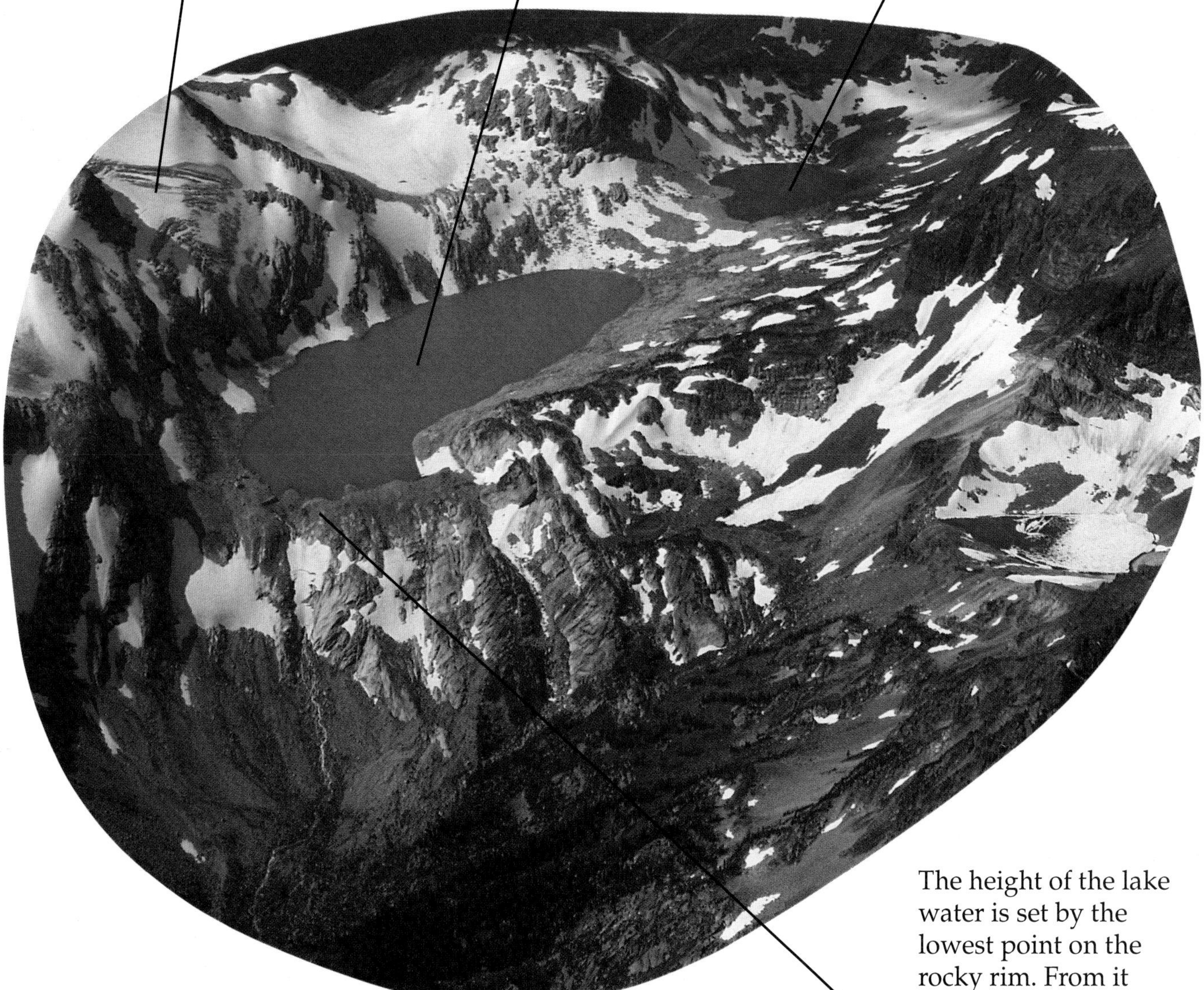

This mountain hollow is still filled with ice, but may form a lake if the ice melts away.

This lake formed at the end of a high-level valley. It is still trapped by a rocky ledge.

This is a true bowl-shaped hollow, or cirque, with a circular lake. It is trapped behind a rock bar that is capped with a rubble (moraine) ridge.

The height of the lake water is set by the lowest point on the rocky rim. From it water cascades down to the main valley.

Bowls of the mountain peaks

The highest mountain lakes were formed in regions that were once the headwaters for streams. Shallow dish-shaped hollows in the hillside became scoured out into deep bowls called **cirques**. When the ice melted away, some of the bowls had rocky rims, while others had rims made from the rubble that was once carried by the ice – known as **moraine**.

For more information on the work of glaciers see the book Glacier *in the Landshapes set.*

Ribbon lakes

The lakes that fill the bottoms of some of the world's most spectacular ice-scoured valleys are called ribbon lakes because they are long and thin.

These ribbon lakes fill valleys that, during the last Ice Age, were cut deeply into the mountain valleys by glaciers. As in the mountain hollows, the scouring was not even; where bands of hard rock occurred less scouring took place but where bands of soft rock occurred deep trenches were worn into the valley bottoms.

When the ice finally melted away this irregular scouring was exposed and depressions filled with water to make ribbon lakes. Some lakes are connected by rivers that flow across the harder rock bands.

Lochs

The word *loch* is Scottish for ribbon lake. The Scottish highlands have many ice-scoured valleys and many famous lakes including Loch Ness (with its supposed monster) and Loch Lomond.

These lakes do not have such steep sides as, for example, those in the Andes (shown in the picture opposite), New Zealand, or Canada, but their long straight shape indicates that they have formed in a valley once scoured by a glacier 'bulldozing' a straight path through the landscape.

Loch Tummel is typical of the lochs for which Scotland is famous.

For more information on the work of glaciers see the book Glacier *in the Landshapes set.*

This deeply trenched valley is called a U-shaped valley because it resembles the shape of a 'U', a clear sign that the valley was created by a glacier.

A string of mountain lakes is sometimes called stairstep or paternoster lakes.

A harder band of rock remains as a ridge, holding back the lake.

This is a ribbon lake, formed in an ice-scoured trench.

The track gives the scale of this spectacular valley in the South American Andes mountains.

Natural dams

There are many natural events that can dam up a valley and cause a lake to form. Sometimes landslides block a valley within minutes (see Thistle Lake, page 10). A flow of lava will also block a valley, causing a dam that steams and hisses as water builds up behind it.

However, ice sheets and glaciers have been the cause of most of the world's natural dams.

Dam spotting
Look at the spot where water flows out of a lake. Is it strewn with boulders like the one shown in the picture on the right or hummocky like the one shown in the picture below? If so, the chances are that it is being held back by a natural rubble and clay dam formed when glaciers melted away.

This view is looking down a valley from the end of a glacier. A small lake has been formed where a ridge of moraine has been left as the glacier has melted back. Even large lakes, such as Lake Como in Italy and Lake Superior in North America can be trapped behind this kind of barrier.

Dams that date from the Ice Age

During the Ice Age, glaciers spread down mountain valleys and pushed out onto nearby lowlands. In the warmer lowlands the ice flow was often matched by melting and the glacier came to a halt.

The end, or snout of a glacier, may melt in the same place for many centuries. As a result huge amounts of debris are brought to the snout and then dropped. Gradually it builds into a broad ridge, called a terminal moraine. Many moraines block valleys and hold back lakes. In the picture on the left a glacier coming from the left stopped where the lake now ends and built a broad low ridge that has since become covered with forest. Look carefully and you will see that a river flows from this ridge to the valley bottom, proving that the lake is held behind a debris ridge and does not fill a rocky hollow.

Salt Lakes

Salt lakes are different from all other lakes because they are dry for long periods and only contain water from time to time. It is therefore most common to find salt lakes in hot, dry environments.

Salt is present in tiny quantities in all river waters, but it is normally flushed away to the sea. However, in places where rivers flow only briefly after storms, lakes rarely overflow. Therefore the more water that flows in to the lake, the more salt that is added.

Between storms the lake steadily dries up through evaporation and the water becomes more and more salty until eventually crystals of salt begin to form as a crust on the lake bed.

This is the salty bed of a dried-up salt lake. It is not smooth, but rough where the crystals of salt have grown.

A salt lake often has water just below the surface even during the driest periods. Here you can see the remains of a lake and the white salty spreads beyond.

The salt crystals that cover the central part of the dried up salt lake bed in the picture below glisten in the sunshine. Near the edges of the lake bed, however, salt is not the most important material and beds are mainly covered with the sand or mud brought into the lake by flooding rivers. Indeed, parts of a lake bed may be so rarely covered with water that there is an opportunity for plants to grow. You can see this in the picture at the bottom of the page.

The largest salt lakes, such as The Great Salt Lake in Utah, USA, and Lake Eyre in South Australia, cover many thousands of square miles.

Crater lakes

From time to time volcanos explode with such violence that they create cracks in the surrounding land. Deep below the surface is the chamber that supplies the lava for the eruption. As the eruption proceeds the chamber begins to empty. If the ground above has been cracked by the violence of the eruption, then there is a danger that the top of the volcano may fall in on itself to make a huge pit, or crater which soon fills with water. This is what has happened in many of the world's crater lakes.

For more information on volcanos see the book Volcano *in the Landshapes set.*

Crater Lake

This is an almost perfectly round lake about 6000 ft above sea level in Oregon, USA.

On the site of the lake a volcano called Mount Mazama once stood. The cone rose to 12,000 ft, but about 6500 years ago the ground was shaken by violent earthquakes and the volcano began to erupt.

On the day of its most violent explosion so much lava was expelled that the feeding chamber was partly emptied, the cone collapsed and blocked the feeding chamber bringing the eruption to a stop. The pit filled with water to make a lake 2000 ft deep.

But Mount Mazama is not finished yet. In the center a new cone is slowly growing. The top now rises above the lake and is called Wizard Island. In time the volcano will erupt again.

Crater lake is over 7 mi across and a journey around the rim stretches over 22 mi.

How crater lakes form

Violent explosions crack the surface rocks.

The lava flows out rapidly during the eruption and leaves a vast empty chamber.

The weight of the volcano causes the cone to fall into the chamber, making a huge natural well that soon fills with water.

A new cone begins to grow in the lake, surfacing as an island.

A crater can be very deep, and the lake that fills it is like a vast circular well.

There is little mud in a crater lake because no rivers flow into it. As a result the water remains crystal clear.

This picture shows a lake that has filled a huge crater in a mountain top on the island of Bali, Indonesia.

Flood plain lakes

During floods rivers spread out vast quantities of debris over the bottoms of valleys. The flat land is called a flood plain, and, because it is so low-lying, water can be found just below the surface. This means that any hollows in the flood plain will usually be full of water, forming mini lakes. Here is how some of these mini lakes are produced.

Ice-made lakes

In regions where winters are extremely harsh the ground has never fully melted since the Ice Age. Where the ice does melt and turn to water, the soil collapses to form hollows which then fill with water to make small lakes. These are often called kettle hole lakes. Canada has many thousands of such mini lakes.

Oxbows and backswamp lakes

Rivers create most of the hollows found on their flood plains. New channels can easily be made and old ones abandoned during a flood because the debris is so easily moved about.

When a channel is abandoned it is no longer flushed clear by running water, and its ends silt up, separating it from the active channel. However, water continues to seep through the banks and keep it full.

Lakes produced in this way are called oxbow lakes. The levels rise and fall with the amount of water in the flood plain, and so oxbow lakes can dry up completely during a drought.

For more information on oxbow lakes and flood plains see the book River *in the Landshapes set.*

This is a billabong in Australia. It is a small lake enclosed by marshy edges. Billabongs are formed in low lying hollows just like oxbow lakes. The name was given to them by the native Australians.

Where water turns to ice it heaves up the ground into mounds, where ice melts to water the ground sinks and makes hollows that fill to make lakes. This hill (known as a pingo) on the flood plain of Canada's Mackenzie River is filled with ice. Surrounding it are lakes formed where ice has melted.

Where land sinks

Rivers and ice alone cannot cut the huge basins that form the world's largest lakes. All of these – some very deep trenches, others broad shallow 'inland seas' – have been formed by the land sinking as a result of the Earth's crust moving.

Troughs and trenches
In some places the Earth's crust breaks into slabs. These slabs can sink down making rift valleys and forming very deep narrow lakes, such as Lake Baikal in Russia and Lakes Nayasa and Tanganyika in Africa.

Basins made from below
Sometimes forces deep within the Earth cause the crust to be pulled down into a wide basin. The basins which hold the salt lakes of Australia (such as Lake Eyre) and the world's largest lakes – the Black Sea and the Caspian Sea – were made by land sinking down in this way.

Nearly all of Africa's lakes were made by Earth movements. This is Lake Nakuru, a shallow lake that occupies a small 'dip' in the Earth's crust; Lake Victoria (Africa's largest lake) and Lake Chad were also formed this way.

The Sea of Gallilee is the northern-most point of the world's largest rift valley – the East African Rift Valley. You can clearly see the steep straight sides of the rift in this picture.

The Great Lakes

Just as a dent can be made in the surface of a balloon by pressing on it, so the huge weight of an ice sheet can push the **crust** down into a broad basin.

During the last Ice Age in North America, the weight of ice sheets was so great that the land sank over the huge area that is now partly filled by the Great Lakes.

The picture above shows how Lakes Erie and Toronto are separated by a slender band of low land. This is being cut away by the St. Lawrence River through the growth of the Niagara Falls gorge (shown in the picture on the right).

The Great Lakes fill a large depression in the crust, but their clear-cut margins are made by ridges of moraine deposited when ice sheets melted away at the end of the Ice Age.

The picture on the left shows part of Lake Michigan and the city of Chicago.

For more information about rift valleys see the book Valley *in the Landshapes set.*

How will the Great Lakes end?

The Great Lakes seem massive and it is hard to imagine that they could ever disappear. But there are two forces ensuring that the lakes will not survive more than a few more thousand years: (i)the barrier, marked by the Niagara Falls will eventually be worn away; and (ii) the land will rise back to where it was before the Ice Age. So, one way or another, the lakes are sure to go.

New words

cirque
a bowl-shaped hollow that is scoured out of a mountainside by a glacier. Cirques begin as stream-cut hollows in mountainsides. In the Ice Age these areas filled with snow and ice. As the ice flowed downhill it cut fastest in the center of the hollow, where it was deepest. After the Ice Age ended many hollows filled with water to make lakes. Many countries have a separate name for the lakes formed in the hollows. One such word for a mountain lake is tarn

crust
the name for all the rocks that make up the solid surface of the Earth. The rocks may seem to be solid and stationary to us, but over a long period of time the rocks of the crust move up and down and also across the surface of the Earth. The forces that make them do this are in the part of the Earth below the crust. This region is molten and turns over and over in much the same way as soup moves when it is heated in a pan

delta
the name for a wedge of sediment that builds up in a lake as a river drops its load of silt in still water. Deltas are similar to icebergs in so far as only a small part of their true size stands above water

Ice Age
the time, beginning about a million years ago, when the world became colder and ice sheets spread across a third of the world's land surface. Lake hollows were made in the early stages of the Ice Age, but they were also made near the end when glaciers could no longer carry their debris and they dropped it in irregular ridges across the land. Ice-formed lakes are common because the ice sheets and glaciers melted away only a few thousand years ago

levee
a ridge of sand and silt that builds up on the edges of a river channel. It is formed during floods when the river spills over its banks. As the fast flowing water spreads over the nearby land it slows down and drops its load, forming a levee

moraine
the name for any material that is carried by a glacier or ice sheet. Most of the material is carried along the bottom of the glacier or ice sheet and is dropped over the land when the ice melts

rift valley
a valley with very straight, steep sides that has been made as a region as the Earth's crust has been pulled apart, allowing blocks of crust to drop

sediment
the name for any material that has been carried in suspension in water and later dropped on the bed of a river or lake. Mud is the smallest size of sediment, silt is a little larger, with sand and finally pebbles as the largest materials

swamp
a region where non-moving shallow water is found and which has been colonised by plants so that, from the air, it looks like land. Swamp is often the name given to wetlands in warm areas and in cooler regions they are commonly referred to as a marsh or bog

Index

MAR 1 ■ 1994